Images
of
San Diego

First Printing October, 1991
Robert D. Shangle, Publisher

Library of Congress Cataloging-in-Publication Data
Images of San Diego
p. cm. ISBN 1-55988-304-9 (soft bound): $6.95
1. San Diego (Calif.) — Description — 1981 — Views. I. LTA Publishing Company.
F869.S22I43 1991 917.94'985.453 — dc20 91-20177 CIP

Production, Concept and Distribution by LTA Publishing Company, Portland, Oregon.
Printed in Thailand. This book produced as the major component of the "World Peace and Understanding" program of Beauty of America Printing Company, Portland, Oregon.

Introduction

"What a beautiful area!" "I want to remember this forever!" "It's absolutely awesome!" "The Creator simply out-did Himself!"

All of these statements are descriptive of the thoughts expressed when viewing this great city, San Diego, that we live in, work in, and play in. And why not. This is a Grand Place.

Images linger in our mind's eye, bringing back those memories of excitement, happiness, family, loved ones, places we've visited, or always dreamed of visiting. One can remember, either because "I've been there," or visited vicariously. We want to hold onto those experiences of "places I've been, things I've done, places I want to see."

The images in this book have been gathered together to assist with those memories and you can give it life. Combining these pictures with your memories make them fill with energy, telling your story that is full of excitement and thrills.

A tribute to San Diego!

San Diego from Point Loma

San Diego Skyline and Harbor Island

Balboa Park

The California Tower in Balboa Park

San Diego Skyline from Coronado

San Diego Skyline at Sunset

Mission San Diego de Alcala

The History Museum in Balboa Park

Mission Bay

Seaport Village

Old Point Loma Lighthouse at Cabrillo National Monument

Spring Flowers Along the Beach at La Jolla

San Diego Convention Center

Mission Beach and Mission Bay

Mission Bay

Ranunculus Flower Field Near Carlsbad

Harbor Drive

Sunset Cliffs

The Old Globe Theatre, Balboa Park

Casa del Prado, Balboa Park

San Diego Downtown Area

San Diego Skyline at Dusk

Mission Bay

Bazaar Del Mundo in "Old Town"

Coronado Coastline

La Jolla

Heritage Park in "Old Town"

Harbor Island

Pacific Beach

San Diego Skyline

San Diego Wild Animal Park

Along The Harbor

University of San Diego Campus

Del Mar

Cabrillo National Monument

Torrey Pines State Park

Sea World at San Diego

San Diego-Coronado Bay Bridge

At the San Diego Zoo

The Serra Museum in "Old Town"

San Diego Downtown Area

Coastline at La Jolla

Hotel Del Coronado and Glorietta Bay

Coastline at La Jolla

Mission Valley

Coronado Cays Area of Coronado

Oceanside

Coronado Golf Course